ENERGY FROM OIL

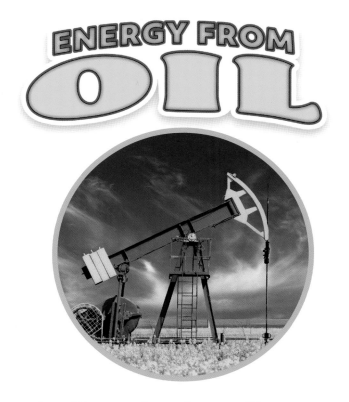

by Karen Latchana Kenney

Consultant: Beth Gambro
Reading Specialist, Yorkville, Illinois

BEARPORT
PUBLISHING

Minneapolis, Minnesota

Teaching Tips

Before Reading

- Look at the cover of the book. Discuss the picture and the title.

- Ask readers to brainstorm a list of what they already know about oil. What can they expect to see in this book?

- Go on a picture walk, looking through the pictures to discuss vocabulary and make predictions about the text.

During Reading

- Read for purpose. Encourage readers to think about oil and energy and the roles they play in our daily lives as they are reading.

- Ask readers to look for the details of the book. What are they learning about oil?

- If readers encounter an unknown word, ask them to look at the sounds in the word. Then, ask them to look at the rest of the page. Are there any clues to help them understand?

After Reading

- Encourage readers to pick a buddy and reread the book together.

- Ask readers to name one reason to use oil and one reason to not use oil. Go back and find the pages that tell about these things.

- Ask readers to write or draw something they learned about energy from oil.

Credits:

Cover and title page, © PhotoStock10/Shutterstock; 3, © mattjeacock/iStock; 5, © kali9/iStock; 7, © Jecapix/iStock; 8–9, © Jubal Harshaw/Shutterstock; 11, © okugawa/iStock; 12–13, © Maximov Denis/Shutterstock, © hanhanpeggy/iStock; 15, © chrishg/iStock; 16, © Sheryl Watson/Shutterstock; 18–19, © milehightraveler/iStock; 20–21, © Miljan Živković/iStock; 22, © WPAINTER-Std/Shutterstock, © SpicyTruffel/iStock; 23BL, © Natnan Srisuwan/iStock; 23BM, © sefa ozel/iStock; 23BR, © mysticenergy/iStock; 23TL, © grandriver/iStock; 23TM, © Andrei Kholmov/Shutterstock; 23TR, © aapsky/iStock

Library of Congress Cataloging-in-Publication Data

Names: Kenney, Karen Latchana, author.
Title: Energy from oil / Karen Latchana Kenney.
Description: Minneapolis, Minnesota : Bearport Publishing Company, [2022] |
Series: Power up with energy! | Includes bibliographical references and
index.
Identifiers: LCCN 2020057398 (print) | LCCN 2020057399 (ebook) | ISBN
9781647478674 (library binding) | ISBN 9781647478742 (paperback) | ISBN
9781647478810 (ebook)
Subjects: LCSH: Petroleum as fuel--Juvenile literature. |
Petroleum--Juvenile literature.
Classification: LCC TP355 .K46 2022 (print) | LCC TP355 (ebook) | DDC
665.7/73--dc23
LC record available at https://lccn.loc.gov/2020057398
LC ebook record available at https://lccn.loc.gov/2020057399

For more information, write to Bearport Publishing, 5357 Penn Avenue South, Minneapolis, MN 55419.
Printed in the United States of America.

Contents

Driving Far

Let's go to the beach.

We need to drive to get there.

What makes our car go?

It uses oil!

4

We need **energy** to make machines work.

Energy gives them power to move.

Gasoline can power cars.

Gas is made from oil.

Oil comes from plants and animals.

They died in oceans long ago.

Over time, they changed into oil.

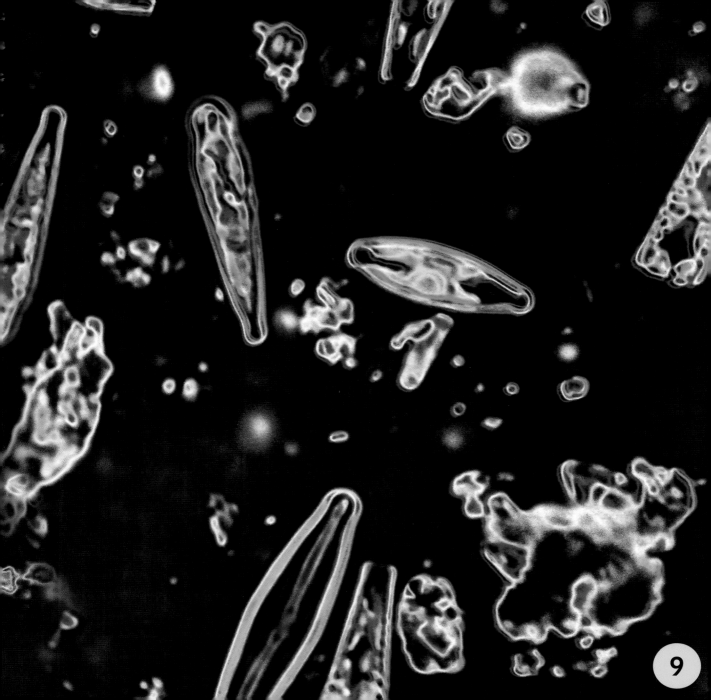

Oil is a dark **liquid**.

It can be found under the ground.

Oil is between rocks.

It is inside some small rocks, too.

11

A large **drill** can dig
a hole to oil.

Oil flows up the hole.

Then, a **pump** moves
oil out of
the hole.

A drill

13

Oil can be made into different things.

It can become gasoline for cars and trucks.

Or it can be made into **fuel** for planes.

Many of us use oil every day.

Lots of machines are made to use oil.

It is easy to find.

But there are some bad things about energy from oil.

It will run out one day.

And oil makes the air dirty when we use it.

Junction I-225   1
Belleview Ave   1¼
Orchard Road   3

For now, we need oil to stay moving.

But we are finding new ways to power machines.

One day, we may not need oil at all.

Energy from Oil

Follow along to see how oil is made.

1 Long ago, plants and animals died. They fell to the bottom of the ocean.

2 Rocks and sand piled up on them and pressed down.

3 The dead things got hot.

4 Over time, the plants and animals turned into oil.

Glossary

drill a large machine that digs deep into the ground

energy power that makes things work

fuel something used as energy to power planes

gasoline a liquid used to power cars

liquid a thing that flows, such as water

pump a machine that moves liquid from one place to another

Index

Read More

Meister, Cari. *Cars (Transportation in my Community).* North Mankato, MN: Pebble, 2019.

Olson, Elsie. *Oil Energy (Earth's Energy Resources).* Minneapolis: Abdo, 2019.

Learn More Online

1. Go to **www.factsurfer.com**
2. Enter "**Oil Energy**" into the search box.
3. Click on the cover of this book to see a list of websites.

About the Author

Karen Latchana Kenney likes biking and reading. She tries to find ways to use less energy every day.